Katie's first riding lesson

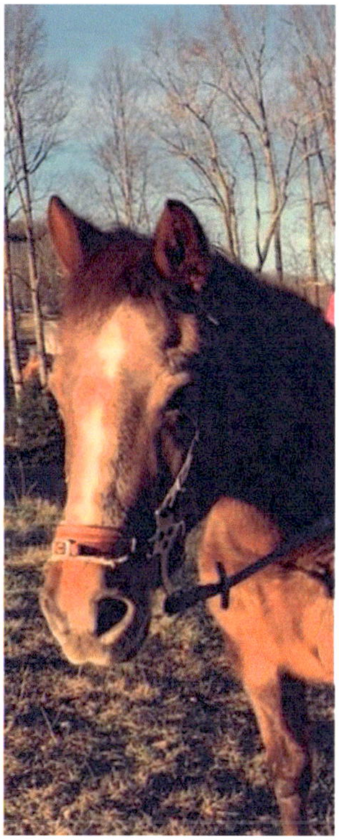

Written by Wendy Cantrell

Told by Sugar

The Cantrell Farm series

This book is dedicated to Sugar and all the wonderful horses that touch our lives.

Diane Ayers with Joshua Mtn Miracle aka “Sugar”

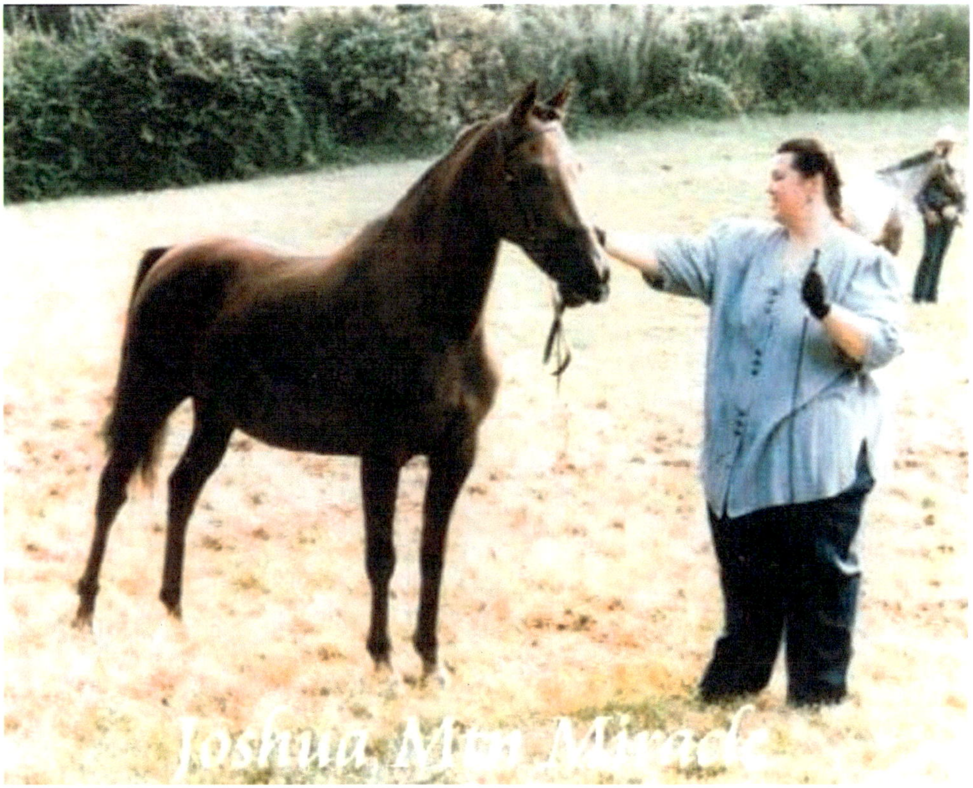

The Cantrell Farm stories will give a greater understanding to the many personalities and traits that make each horse an individual. They too can be happy, sad, confident, and fearful or any of the feelings and emotions that humans have. After years of instructing others how to ride and care for horses, horses never cease to amaze me.

Available from Amazon, Kindle and other retailers

IBSN-10:1-7325057-0-5

Katie's First Riding Lesson

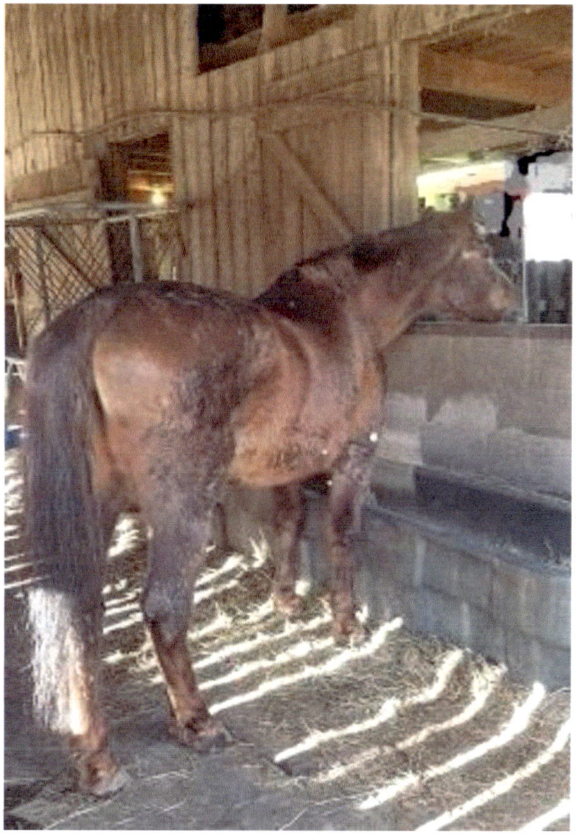

Hello! Anyone coming to see me today? Some treats would be nice.

Maybe brush me?

My name is Sugar. As you can see, my coat color is a seal brown or liver chestnut. I am a Morgan horse. The past sixteen years I have helped many little people learn how to ride. Before that I was a brood mare. A brood mare is a horse that has babies. I have four children, Joy, Stormy, Windy, and Lightning.

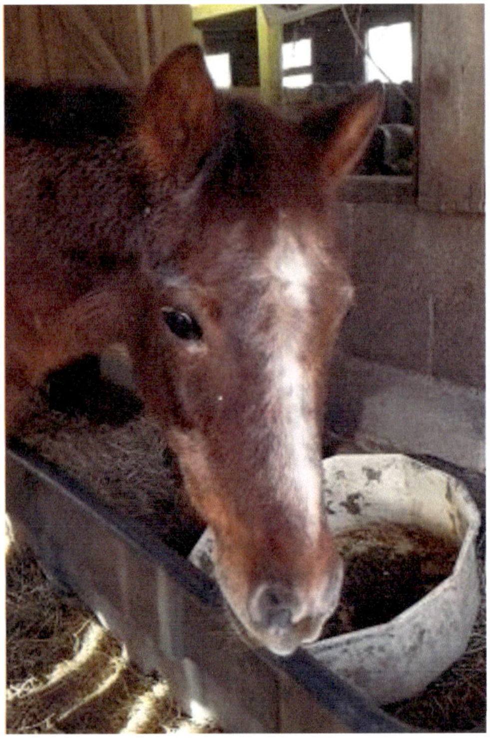

I live at Cantrell Farm with my friend Cake. I share a paddock or room with Cake. We share the hay but not the grain. I get more grain than Cake does, so I eat in peace and quiet out of my big bowl. Cake loves to lick my bowl.

I like looking through my window to see what is going on in the barn. I see you.

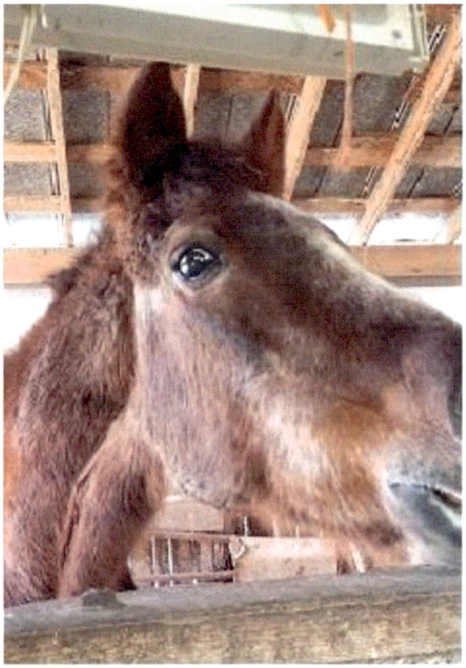

I've been helping people learn about horses for many years. I could tell you lots of stories about people I've helped. But today I want to tell you about Katie's first lesson. Have you ever wanted to learn about horses? We are really cool with a great sense of smell and keen eyesight. Katie is eight years old and does not weigh very much. She is my kind of kid, gentle, calm and speaks to me in a quiet voice. Did I tell you we have excellent hearing? We do. I can hear the back door open at the house which is about fifty yards away from the barn and our paddock is on the opposite side of the barn! I will call to Wendy, our main mom that feeds us, to hurry up with the food. I can hear her mutter, "How does she know I opened the door?" Told you, we have excellent hearing.

Wendy feeds us hay first. There are six other horses here on the farm. We all help teach people about us and how to ride and care for us. Sometimes we go out in the pasture which I like very much.

Back to Katie. Her first lesson was very warm for a winter day and I have my winter coat. When it gets this warm I like to lay down in the cool dirt. This helps me cool down but it also makes me very dirty. That's OK, I have Katie to brush me today.

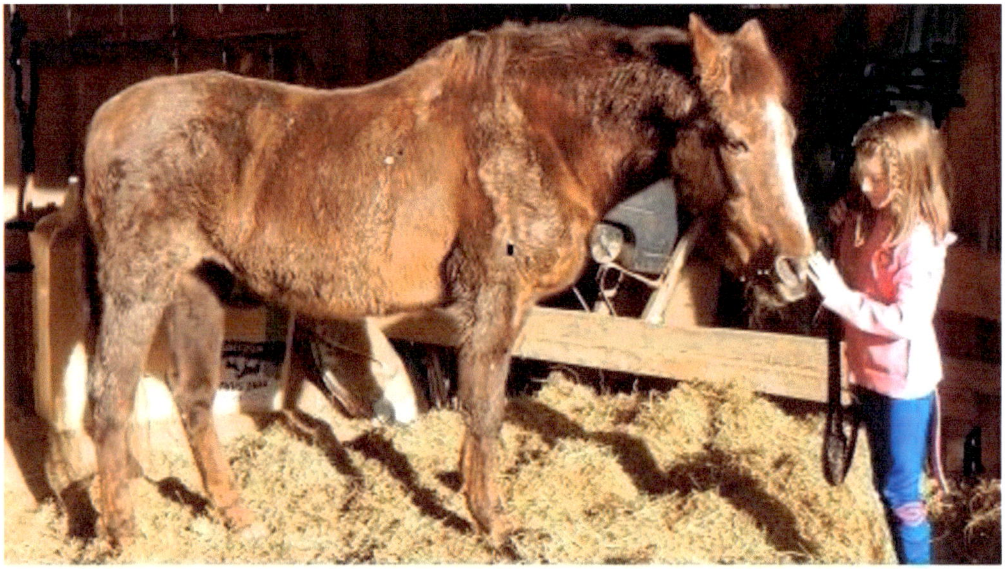

I want Katie to let me smell her hand and get to know me. She held her hand out, palm side down so I could smell the top of her hand. That way I will recognize her when she comes back. Palm side up means you might have a treat. I'm so glad Katie didn't pet my nose. It is bad manners.

I think Katie was a little afraid of me at first because I am so big. I weigh about 880 pounds. I measure 13.3 hands tall. A hand is how people measure how tall horses we are. A hand is the same as four inches. People measure how tall a horse is at the withers. My withers are where my neck and back meet at the end of my mane. The mane is the long hair on my neck that people like to play with.

Wow I have a lot to explain. Learning to ride is learning about me.

So now Katie learns how to put my pink halter on. A halter is used to lead and groom us. I try to help her by putting my nose in the right spot. You see, if Katie stands on my left and holds the buckle in the left hand and the strap that goes behind my ears in her right hand, I can see the opening to put my nose in. She just has to buckle the strap. I hear Katie ask Mrs. Wendy "What hole do I buckle the strap in?" Great question Katie. If Katie puts a finger between my cheek bone and the nose band, it works best and feels just right. That way it is not hanging down on my nose, too low, or rubbing my cheek bone from being too high.

Next Katie has to learn how to lead me. The best way to lead me is to hold the lead rope and to walk next to me. She always needs to go through the door first. I will follow her.

Katie has her right hand about twelve inches below my halter. The rest of the lead rope is folded in her left hand. She can give me better directions if she holds the lead rope this way.

She also had to learn the proper place and way to tie me to be groomed. Horses are very strong and can pull boards off fences. I knew a horse that got scared and pulled a post out of the ground. We normally do not pull back unless we are scared. I get groomed in my stall where I feel safe.

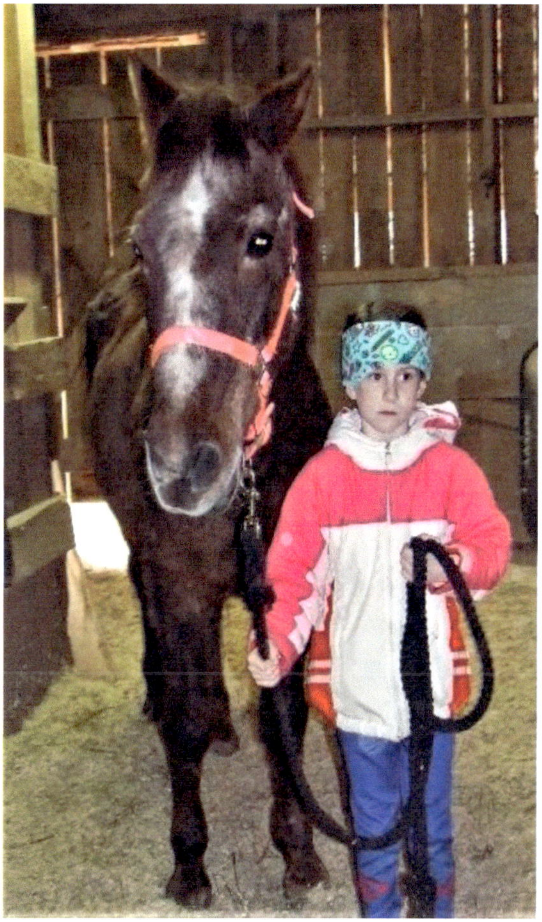

Katie uses a slip knot so if I need to be untied quickly she can pull the end of the rope and the knot comes out fast. I am trying to tell you how to do this right so no one gets hurt.

To make a slip knot, make an "e" with the end of the rope (Not the part attached to me). Place the "e" on top of the rope going to me.

With the end of the "e" make a bunny ear, go under the rope going to me and back through the "e". Just the bunny ear shows.

After Katie has me safely tied in my stall, she has to learn about which grooming tools to use. Grooming helps us learn about each other. Since she grooms even the clean area, she can check for any booboos, places I might have gotten hurt, and of course make me clean. I like being clean, it makes me feel like doing my best.

Katie uses a towel for my face. While she wipes my face she checks to make sure nothing is wrong with my eyes. She holds the halter as she wipes my face in the direction the hair grows. If my nose is dirty, she wipes it too with a paper towel. A curry comb is used to loosen dirt and hair on my neck, legs and body. She uses a body brush to wipe the dirt and hair away brushing in the direction the hair grows and a hoof pick to clean my feet.

Sometimes when I get really dirty, Katie will have to use a vacuum or blower to get the dirt off.

Katie starts cleaning my face, then goes to my neck, chest, shoulder, back, barrel, belly, croup (the top of me near the tail), thigh and legs on one side. Then does the same on the other side. Finally she cleans my mane, tail and hooves.

I have the same body parts as you. Some have different names though. I have one finger nail or hoof but you have five. I also have bones in my tail. I do not have a collar bone but you do.

My favorite body brush that Katie uses is one with natural bristles. This kind is not as colorful but they don't shock me.

Body brushes that are made of nylon bristles can cause static electricity when used to brush, they shock me.

There are many different types of brushes. A stiff bristle for the mane and tail plus winter mud when my hair is long. A really soft brush for tender parts like my belly, ears and inside thighs. A plastic brush for washing dirt off the bottom of the legs when warm enough to wash. A grooming stone and shedding blade for spring when my new short summer hair comes in and my long winter hair is falling out are some of the other tools Katie will learn to use.

Here Katie is holding a natural bristle body brush and gel curry.

Katie has to learn about the three main parts of my fingernail. Oh me, fingernails are called a hoof. My hoof has a hoof wall, a sole, and a frog. When Katie picks the dirt and rocks out of my hooves, she holds onto the hoof wall. My sole is on the bottom, and the triangular part in the middle is called the frog.

This is the bottom of my hoof before it is picked showing the sole and frog. I can get stones stuck next to my frog. These stones can make it hurt when I walk. Katie also cleans out bacteria in my hoof that causes thrush. Thrush can eat my frog up.

It does not bother people but it will give me a sore foot.

Katie stands next to me while she is picking my hooves. If she does not stand next to me and pulls my leg out, I will lose my balance.

Now I am ready to be saddled. For Katie's lesson she will be riding with a bareback pad; that means she won't have stirrups. Katie will learn her balance so much faster since she will have no stirrups.

First she places a thin, pink saddle pad on, then a black, shock absorbing pad, finally the bareback pad. Don't I look good in pink, it's my favorite color! The bareback pad has a girth to hold it on. Katie gets it snug but not tight. Balance keeps her on my back not the pad. Good job remembering your helmet Katie, it will keep you safe.

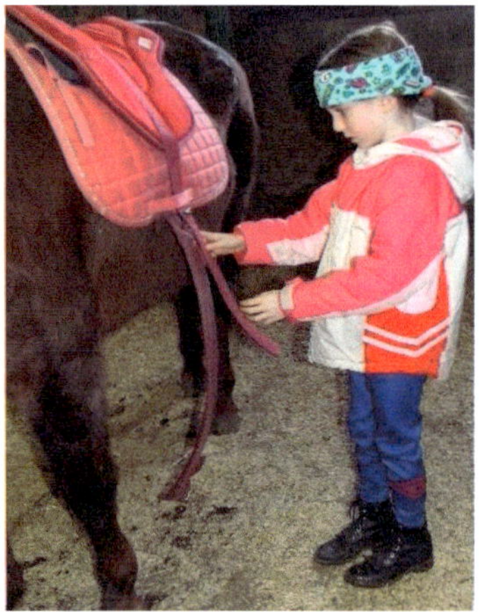

My bridle is next. My bridle has a hackamore nose piece so I don't have a bit in my mouth.

Horse's faces are tender just like yours, so Katie is careful putting it on.

After the reins have been put over my ears and on my neck, Katie gets the nose band on followed by pulling the crown piece (top of bridle) over one ear, carefully folding my ear forward, then the other ear.

No one likes having their ears crunched and this way the cheek piece does not drag over my eye ball.

Katie checks to make sure that the bridle is in the right place. The crown piece goes behind my ears, the brow band is on my forehead right below my ears and the noseband is where my halter was. Then buckle the throat latch. I like that Katie is gentle.

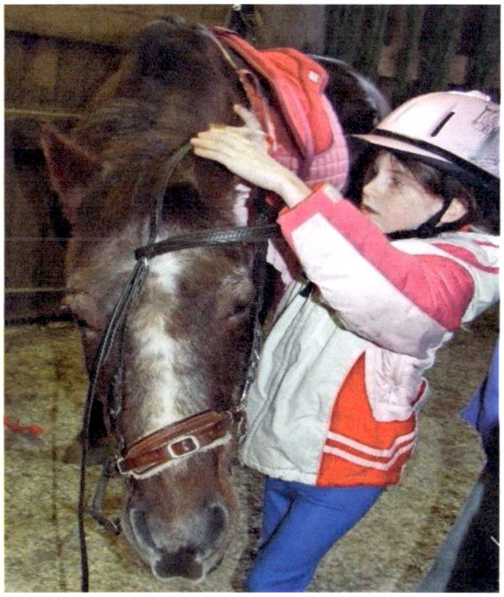

Katie removes the reins from my neck to lead me up to the ring where we will play. I like that she is walking beside me, holding the reins so they do not drag on the ground.

Katie has to learn how to ask me to turn right and left, to stop and to back up while beside me on the ground. This is called ground work. Katie needs to learn how to communicate with me on the ground before she rides.

This is where a lot of people get confused. Some people think they "control" us, but it is really learning how to communicate with us. We have to learn to speak the same language.

Horses get confused when they are given directions they do not understand. My friend, Callie, a black and white paint horse told me about a rider that rode her one day. The rider kept swinging her legs. Callie thought they wanted to go faster, but they didn't. She was so confused that she shook her head. The rider got scared, jumped off and started jerking the reins which scared Callie. It was a mess!

I have learned to follow the angle of Katie's body. That means which direction her body is facing is the direction I need to go. If Katie leans forward she wants me to go or if she leans back I know she wants me to slow down. When she stops moving her feet, my feet stop too. Since horses communicate mostly with body language, I watch what she does with her body to know what to do.

When she rides, I feel for her legs and body. Pressure from legs or body are signals to me.

People do not have to kick their legs for us to go. Mrs. Wendy taught me to go when my rider's legs are gently squeezed, that is all that I need. It depends on what we are taught, just like you. I can explain this in more detail when Katie takes another lesson.

Katie gets a leg up meaning someone helps her on. I wait patiently while Katie gets directions on how to hold the reins, how to sit and what to do with her legs.

Riders stay on using balance not hanging on as most people think.

I can hear Mrs. Wendy explain to Katie how to hold the reins the right way. Her hands stay quiet near my withers. (The withers are

where the mane ends). Katie holds the reins with her thumb up so she just needs her fingers to cue with. Heels are down and sits proud, not like a bag of potatoes.

Katie is almost ready to move but she needs to know the cues (signals) first.

To go, a gentle squeeze of her legs tells me to walk.   When I walk, Katie stops squeezing so I know I have done what she wanted.

To turn to the right, she presses only her left leg against my sides pushing me right. She also needs to shorten the right rein just a little and squeeze her fingers of her right hand that turns my nose to the right.

To turn to the left, Katie presses her right leg against me but not her heel to push me left. Shorten the left rein and gently squeeze her left fingers. Simple as that and I turn to the left.

The last cue Katie needs to learn for her first lesson is how to stop by breathing out, while not letting her legs swing or bump my sides. If I feel Katie not moving, I stop. I also know the word “whoa”. Now sometimes if I get excited she might have to give a gentle pull on one rein to get my attention. Of course some horses are not good listeners like me.

Some horses have not been taught signals.

Katie is using her left rein to tip my nose to the left. I can feel her right leg pressing so I know to go left. When she pushes with her leg she does it in rhythm with my walk. This makes it easier for me to understand. By keeping her hands near my withers, she is able to balance. If Katie's hands bounced around, I would not know if she was giving me a signal. If her legs bounced I would get really confused about which signal she is trying to give me.

Now Ms. Wendy put me on the lunge line. A lunge line is a flat rope attached to my bridle that she holds. I can move in a circle with directions from Ms. Wendy. This allows Katie to learn about balance.

We are going to stop for a minute so you can meet Cake. Cake lives in the same paddock with me. We are good buddies. I asked Cake how she thought Katie was doing.

Once Katie got the hang of stop, go and turning, we get to play the ball game.   Cake wanted to do the ball game but she had to help her rider learn how to keep a steady rhythm at a trot instead.

Katie gets a ball off the barrel and rides down to the bucket to drop it in. Katie has to communicate with me to get close enough to the barrel so she can get a ball. After she has the ball we ride to the green bucket on the ground at the other end of the ring to drop the ball in. Once I realized what game we are playing it was easier for me to know where to go.

Sometimes Katie would swing her legs, accidently bumping my sides. I thought she was giving me a signal to go. Mrs. Wendy would tell Katie to keep her legs still. With practice she will get it.

But it is time to get off and take care of me. Katie holds on to my mane and slides off while still holding the mane, her feet touching the ground first. It doesn't hurt when she holds my mane and it helps Katie to get down safely, so I don't mind.

Taking the reins over my head, she leads me back to the barn but not before I get a cookie.

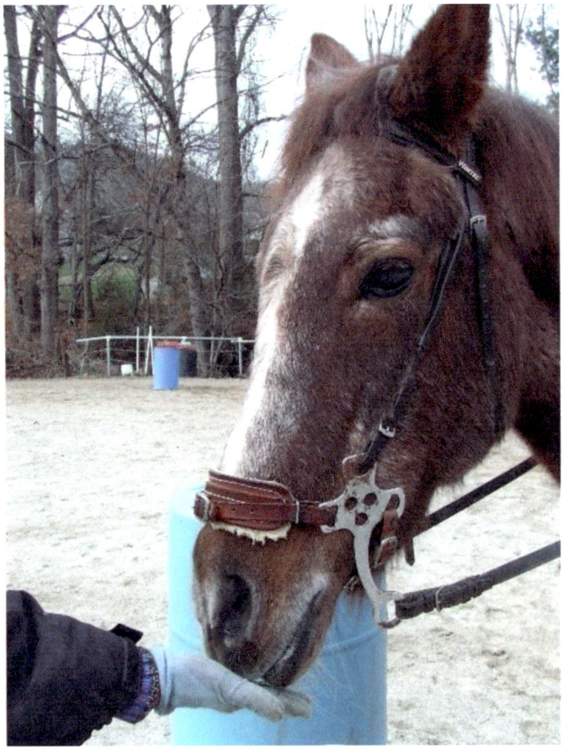

Cookies are one of my favorite treats. I know I have done a good job when it is cookie time! Ms. Wendy holds her hand flat so I can have my cookie. I use my lips to take the cookie out of Mrs. Wendy's hand. It is so yummy!

Katie walks beside me to the barn holding the reins in both hands; so we don't trip or step on them.

Once in the barn Katie unbuckles my throat latch and slips the bridle off and puts my halter on then holds the top of the bridle to carry it.

Katie puts my bridle away, then comes back to take off my saddle pads. She makes sure the saddle pads are clean and dry before putting them away. If they are not clean when she puts them away they won't be ready for our next lesson.

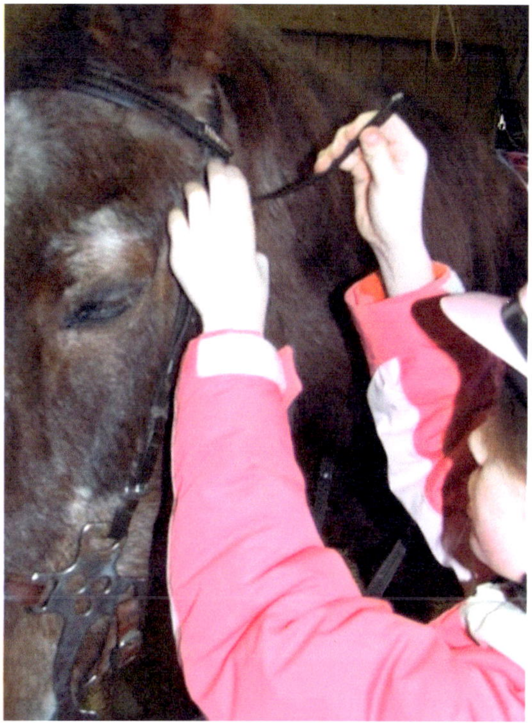

The bridles are hung up by the crown piece with the reins positioned on top. All these bridles have hackamores not bits for communication.

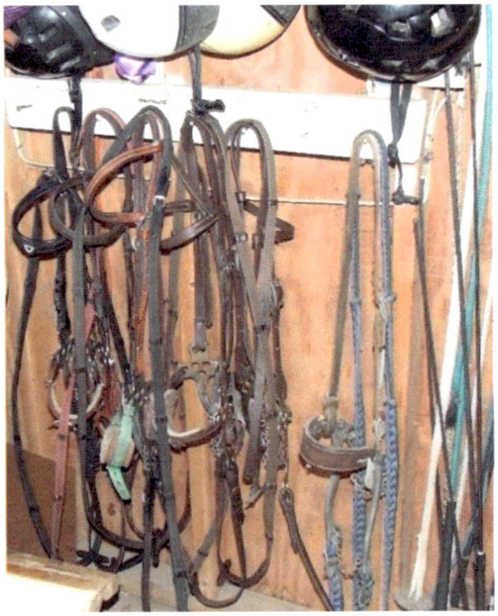

Katie brushes my back, belly, legs and around my ears while I eat. My grain tastes yummy after all the hard work. Sometimes I get sweaty during lessons. If I get sweaty in winter, I would get cold. So I would need a light weight blanket called a cooler to wear while I am dried off.

Since it is winter time, I have a blanket to wear when it is cold. Katie buckles the front straps first before she unfolds the blanket.

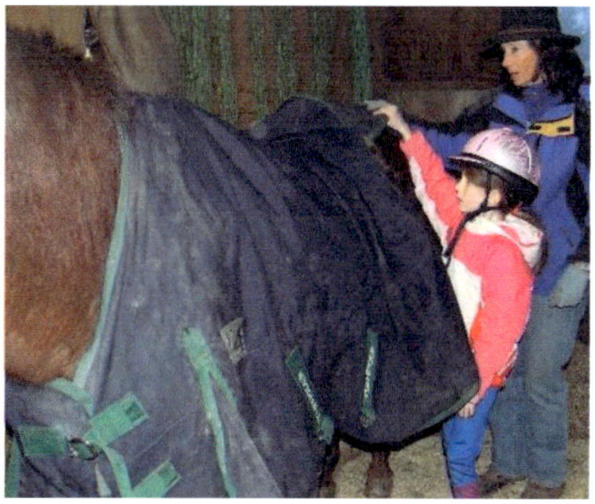

Then hook the belly straps which criss cross under my belly.

I also have a thick, pink blanket but it is too warm for tonight.

I finish my food while she cleans up my manure (poop) and fills my water. I poop ten to fourteen times a day. If my manure is too hard it might mean I am dehydrated and need more water to drink. If too soft or too few piles, I might be getting sick so it is always good idea to check the manure.

Also in the winter the poop can freeze making it hard to walk on, like walking on big rocks. So I appreciate Katie picking it up. Keeping my feet clean and dry prevents hoof problems that are painful and costly.

Katie, I am done with my grain. You can get my bucket. Wait, let me double check it, I might have missed a bite.

Now that I finished my grain, I am ready to go eat my hay. I hope Katie has put it out. I eat about twenty pounds of hay a day or roughly two percent of my body weight. Hay is dried grass. In the winter when the grass does not grow and when it is cold I eat more hay. Katie is bringing me hay that is not dusty or moldy or full of weeds. I can smell the hay that Katie put out. It smells good and is still green. I can't wait!

Katie leads me back out to my paddock that I share with Cake and takes off my halter. Hurry up Katie, Cake might eat my hay!

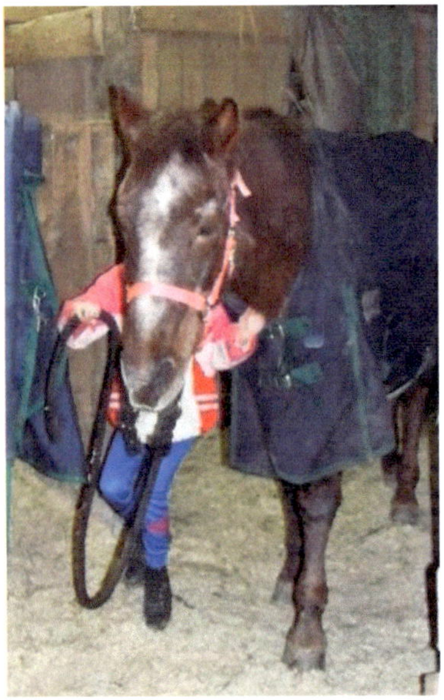

Katie puts away my halter, the brushes and any other tools she has used.

Finally, Katie fills up my water. I am so thirsty. I can drink about ten gallons of water in a day.

Katie's first lesson was really good. I like Katie. She is quiet, gentle and kind.

I know I called it a riding lesson, but it really is a caring lesson. See you next lesson.

Made in the USA
Monee, IL
05 November 2026

67378647R00021